Engineering  Wonders

CRUISE SHIPS

W9-BIJ-034

Kaitlyn Duling

Rourke
Educational Media
rourkeeducationalmedia.com

Before, During, and After Reading Activities

Before Reading: Building Background Knowledge and Academic Vocabulary

"Before Reading" strategies activate prior knowledge and set a purpose for reading. Before reading a book, it is important to tap into what your child or students already know about the topic. This will help them develop their vocabulary and increase their reading comprehension.

Questions and activities to build background knowledge:
1. *Look at the cover of the book. What will this book be about?*
2. *What do you already know about the topic?*
3. *Let's study the Table of Contents. What will you learn about in the book's chapters?*
4. *What would you like to learn about this topic? Do you think you might learn about it from this book? Why or why not?*

Building Academic Vocabulary

Building academic vocabulary is critical to understanding subject content.
Assist your child or students to gain meaning of the following vocabulary words.
Content Area Vocabulary
Read the list. What do these words mean?

- *architects*
- *behemoth*
- *buoyancy*
- *dense*
- *diesel*
- *geolocators*
- *noncombustible*
- *pistons*
- *propel*
- *propeller*
- *turbines*
- *welders*

During Reading: Writing Component

"During Reading" strategies help to make connections, monitor understanding, generate questions, and stay focused.
1. *While reading, write in your reading journal any questions you have or anything you do not understand.*
2. *After completing each chapter, write a summary of the chapter in your reading journal.*
3. *While reading, make connections with the text and write them in your reading journal.*
   a) *Text to Self – What does this remind me of in my life? What were my feelings when I read this?*
   b) *Text to Text – What does this remind me of in another book I've read? How is this different from other books I've read?*
   c) *Text to World – What does this remind me of in the real world? Have I heard about this before? (News, current events, school, etc....)*

After Reading: Comprehension and Extension Activity

"After Reading" strategies provide an opportunity to summarize, question, reflect, discuss, and respond to text. After reading the book, work on the following questions with your child or students to check their level of reading comprehension and content mastery.
1. *What are the most important things an engineer needs to consider when designing a cruise ship? (Summarize)*
2. *Why are there so few companies in the cruise ship building business today? (Infer)*
3. *What role does density play when it comes to keeping a ship afloat? (Asking Questions)*
4. *If you were going to design a luxury cruise ship, how would you make sure it stayed afloat, and how would you attract guests? (Text to Self Connection)*

Extension Activity
Build your own cruise ship! Using aluminum foil, create a boat that can float in a tub of water. Will it have propellers? Should it be heavy or lightweight? How will you make it stay upright? After you're finished designing and building, place pennies on your ship—just one coin at a time. See how many it can hold before it sinks!

TABLE OF CONTENTS

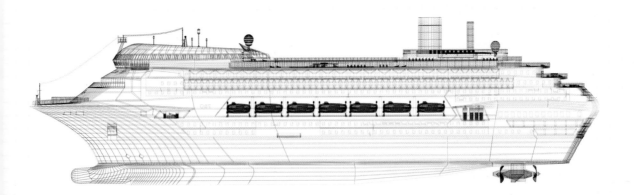

SETTING SAIL

Your feet are planted on the floor and your hands are holding a guardrail. As you let your gaze take in the blue sky above, the endless line of the ocean enters your vision. You hear the "BUHHRR BUHHRR" of horns, see everyone waving and smiling, and suddenly you feel movement. You are no longer on land—you have set sail, whisked away on one of the world's largest ships!

What can you see from the deck of a cruise ship? Whales, dolphins, birds, and other wildlife might be playing nearby.

Luckily, this is a cruise ship. It has everything you might need, including beds, restaurants, bathrooms, and lifeboats, just in case there is an emergency. Over the past few decades, cruise ships have become so safe and so popular that you should not have to worry about any accidents. You and the thousands of other passengers on the ship will be able to cruise peacefully from port to port without a care in the world.

On a cruise ship, the rooms are called "cabins" or "staterooms." They are small, but can have amazing views of the water!

As the years go by, cruises continue gaining popularity. People of all ages and interests decide to climb aboard these huge ocean liners for trips on the high seas. Families, couples, and single adults all love to take cruises.

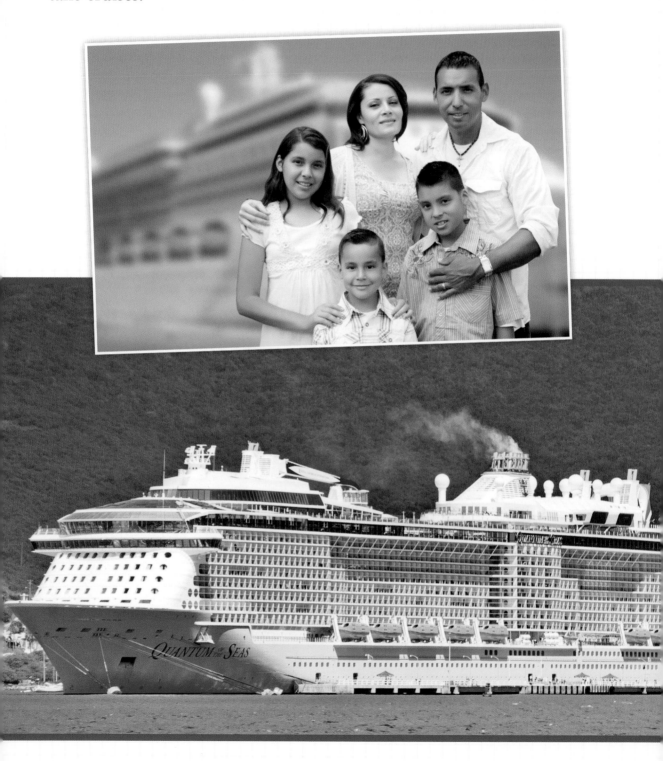

After boarding, passengers can enjoy magnificent meals, professional shows, pools, spas, and even waterslides on the ship! You can find cruise routes around the world, with destinations from chilly Alaska to the warm Caribbean beaches. Some cruises are short, just three or four days long, while others extend to one or more weeks.

Cruise ports often hold more than one ship. These days, cruising is very popular!

A cruise vacation is more than just a getaway for you and a few friends. These **behemoth** ships can fit thousands of passengers, not to mention the thousands of crew members that will be aboard at the same time.

On a cruise ship, the pools are on the uppermost decks so that swimmers can get as much sunshine as possible!

With all of those people, along with their luggage and food, how on Earth can one of these ships stay afloat? Cruise ships are masterpieces of engineering. Their main goal? Don't sink!

Symphony of the Seas
The largest cruise ship in the world is the Symphony of the Seas, operated by Royal Caribbean. The Symphony can hold more than 6,600 people, including 2,200 crew members. Its 1,188 foot (362 meter) frame includes 18 decks, 24 pools, and a tropical park with live plants.

DESIGN TIME

You may have heard of popular cruise lines such as Carnival, Royal Caribbean, and Norwegian. Each line owns several ships of varying sizes, and they all visit a variety of destinations.

The cruise operators don't design these ships. They order them from specialty ship designers. Why? It takes expert knowledge to design a huge, floating vessel.

Some cruise ships, like the Wyland, *are decorated with bright designs to help set them apart.*

The Regal Princess *has a classic cruise ship shape and design.*

This ship in the Meyer Werft building port is almost ready to set sail.

The Fincantieri building port is a busy place.

Most cruise ship design and construction is done by one of a few major companies. Meyer Werft, Fincantieri, and GEM are just three companies that are known for their design and shipbuilding, which is done in dedicated coastal ports.

Each new ship will employee about 20 to 30 designers, so no one person should expect that the finished ship will look the same as their designs. It is truly a team effort! There are **architects**, engineers, and interior designers on the team, each working on a different aspect of the ship.

Sometimes a short ceremony is held to mark the laying of the keel in a new cruise ship.

Cruise ship builders and engineers get to have fun sometimes! These workers got the chance to test out a go-kart track on the Norwegian Bliss.

A designer needs to consider factors such as safety, comfort, fun, and floatability. New cruise ships are introduced each and every year, so designers always need to be thinking about the "next big thing." In this case, it is often the "next biggest thing," as cruise ships continue to grow in size and scope!

High-Tech Cruising
Cruise ships continue to introduce new technology. Recently, some ships have adopted wristbands that passengers use to open their cabin doors, order food, make purchases, and sync with the cruise app. The wristbands can also act like **geolocators** to help parents keep track of their children onboard.

It takes a whole team of designers to plan a giant cruise ship.

These days, designers are able to employ computer programs to help them work on new ships. These programs allow designers to think about the inside of the ship, the outside, and everywhere in-between.

There are more than 300 cruise ships making their way around the world's oceans, seas, and rivers at any given time. Each year, more than 25 million passengers board those ships! Cruise designers and engineers must work diligently to make sure those millions of people stay safe and happy while at sea.

Though these ships are all run by different cruise lines, they dock next to each other in Nassau, Bahamas.

BUILDING A CRUISE SHIP

Ashipbuilder is also known as a shipwright. So, how do shipwrights construct huge ocean liners? Do they build them on the water? Not a chance! It takes about 12 to 18 months to build a cruise ship, and nearly all of the ships that are built come to life in one of four shipyards.

Indoor spaces, like this one at the Meyer Werft shipyard in Papenburg, Germany, are the first stop on a new cruise ship's journey to completion.

Cranes tower over Fincantieri shipyard in Genoa, Italy.

Japan, Germany, Finland, and Italy are home to the shipyards where most cruise ships are constructed. While these shipyards are located near the water, the ships themselves are built on land before being transferred to a river or ocean.

You might think that a cruise ship is built from the ground up, but these ships are actually built upside-down! This helps **welders** construct the steel components of the ships, which are finished in sections. When finished, the sections are welded together and engineers check to make sure the ship is watertight.

Welding is a dangerous, fiery job. Welders must wear protective gear in order to prevent burns.

Cruise Ship Costs

Depending on the size and scale of the ship, a new ocean liner can cost hundreds of millions of dollars to billions of dollars to design and build. Companies order the ships several years in advance in order to plan for the future and give the builders time to complete the massive orders.

The rooms where guests stay while on a cruise ship are called staterooms. Most of the staterooms are made off-site and then hoisted into the ship with an extra-large crane. Picture a hotel room that is made in a factory and then inserted into a building—a building that floats! Each cruise ship contains thousands of identical staterooms. Some have balconies. Others don't have any windows at all.

Ship parts are made off-site to speed the process along. Building the rooms and sections off-site also ensures that there is enough room in the shipyard for all the workers, machines, and supplies necessary for shipbuilding. Once a component is complete, it is flipped over and settled, right-side-up, into a dry dock. This is an area near the water that has floodgates at one end.

This ship, the Le Boreal, *was nearing completion while still on dry land. It was put into service in May 2010.*

Once the whole ship is finished, the floodgates are opened and the ship is ready to float.

Archimedes' Principle
The law of **buoyancy** is also known as "Archimedes' principle." Archimedes was an ancient Greek inventor who discovered buoyancy. The principle tells us that a ship, when launched, sinks into the ocean until the weight of the water it displaces is equal to its own weight.

4 lb

1 lb

0 lb

3 lb

ENGINEERING A CRUISE SHIP

W hen designing and building a ship of immense size,
what do engineers need to consider?

This ship, the Le Boreal, *suffered a major fire in November 2015 while it was traveling in the Falkland Islands. All 347 passengers and crew had to board life rafts and abandon the ship, which had lost all power.*

One of the most important factors is safety. Because a ship sails far away from land, it must be protected against fire.

Jobs in Engineering

An engineer is a person who invents, designs, tests, and builds. They create products, design roads, test airplanes, and build cruise ships, among other things. Engineers often specialize in one area, such as aerospace engineering, mechanical engineering, or civil engineering.

In order to solve this problem, there is virtually no wood used in the design of a ship. Floors, walls, and even furniture are made of **noncombustible** materials such as metal or ceramics.

Royal Princess

Weight is another key consideration. The weight of the ship is also known as the burden. The average weight of a cruise ship can be anywhere from 20,000 tons (18,144 metric tons) to more than 60,000 tons (54,431 metric tons). Ships keep getting bigger and bigger every year!

Safety First
To keep passengers safe, ships hold mandatory lifeboat drills. The ships are outfitted with sprinkler systems, life vests, smoke detectors, and a security team. Safety within the cruise industry is closely monitored by the International Maritime Organization.

These ships can be heavy as long as they are also buoyant. Buoyancy describes the way in which water pushes up on the cruise ship, helping it float.

As the ship moves through the ocean, it pushes water away, which then pushes up on the ship—keeping it up.

Allure of the Seas, *Third largest cruise ship*
Weight: 225,282 tons (204,372 metric tons)
Length: 1,187 feet (362 meters)
Staterooms: 2,742
Passengers Max: 6,687

Engineers also need to make sure that a well-designed cruise ship is less **dense** than the water around it. A bowling ball is dense, while a beach ball or ping-pong ball is not.

Does a bowling ball float? Not a chance! It would sink right to the bottom of a swimming pool.

In January 2012, the Costa Concordia *struck a rock in the Tyrrhenian Sea, tearing open a gash, flooding the engine room, and eventually causing the ship to roll onto its side. Over 4,000 passengers and crew were evacuated in a six-hour period. Thirty-two people died.*

To ensure that a cruise ship doesn't sink, engineers are careful to consider the amount of open space on a ship. These are areas that aren't filled with restaurants, movie theaters, engines, and crew cabins.

Engineers also make sure that the ship is built with lightweight, strong materials like aluminum and steel.

A Floating Theme Park
On a cruise ship, you can find zip lines, water slides, golf courses, water shows, parks, robots, and so much more. Engineers now have to work strategically to include these elements.

Today's cruise ships are full of attractions.
This one features multiple water slides,
pools, and even a high ropes course.

31

CRUISING THROUGH TIME

Today's advanced technologies and materials weren't always available to shipbuilders. The first cruise ships were constructed in the mid-1800s. The first sightseeing cruise set sail in 1857.

Ships like this one, the RMS Strathaird, *built in the early 20th century, were some of the first to offer luxury cruising.*

In the 1800s, cross-ocean journeys finally became possible, in part because of the steam engine, which allowed ships to **propel** themselves across vast expanses. Through the decades, engineers constructed ever-more powerful and efficient engines.

The Prinzessin Victoria Luise *of Germany was the very first vessel built exclusively for cruising. It was completed in 1900.*

Oasis of the Seas

The last steam-engine propelled cruise ships were retired in the 1980s. The majority of today's ships are powered by **diesel** engines, just like the engines used to power semi-trucks and some automobiles—but much, much bigger!

Just like the first cruise ships, today's behemoths hold their engines in the very lowest decks, which increases stability.

If the center of gravity is kept as low as possible, it is easier and safer for a ship to cruise along through the ocean.

And how does it push itself through the water? With a number of parts that work together. The diesel engine in a ship moves **pistons** up and down, which turn a crankshaft.

The crankshaft is connected to a **propeller** shaft, which turns a propeller. The propeller churns up the water, literally propelling a ship to move forward or backward.

A cruise ship takes upkeep. Here, a crew replaces the propeller shafts on a ship while it's in port.

Since the first ocean liners were built, cruise ships have been home to luxury. Traveling on a cruise ship should feel like staying in a hotel, with all of the comforts and capacity of a city.

Some cruise ships, like the Viking Star, *are particularly luxurious.*

The Titanic
When it set sail in 1912, the Titanic *was the largest manmade moving object in the world. When it sank, about 1,500 people lost their lives. This happened despite the ship carrying the legal number of lifeboats.*

With thousands of passengers, engineers have to figure out how to supply electricity to the entire ship. How is this done? Through a huge generator, kept in the lower decks, along with an elaborate distribution system.

Cruise ship engineers are not only interested in making ships beautiful. They also have to consider the power, water, sewage, and safety needed on a ship.

At night, a cruise ship lights up! All of that electricity takes a lot of planning and careful engineering.

INNOVATIVE ENGINEERS

Cruise ships have come a long way since the 1800s. They continue to grow in size, thanks to innovations in propulsion technology. Many propellers are now installed in pods. This allows them to make sharper, smaller turns, even when the ships themselves are long and wide.

Today's ships can make slight movements, allowing them to avoid choppy waters and icebergs, and helping them to sail smoothly into port.

Speedy Ships
The average cruise ship travels at about 23 to 27 miles (37 to 43 kilometers) per hour. High-speed ships can go as fast as 34 miles (54 kilometers) per hour.

The ships of tomorrow continue to push the boundaries of luxury. More passengers, activities, and amenities require more power. The latest ships come equipped with gas **turbines** and diesel electric engines that are more efficient and powerful than the diesel engines and steam engines of yesteryear. This ensures that the lights won't go out in the middle of an ocean voyage.

As cruise ships continue to grow in size and scope, we can count on ever-more impressive feats of design and engineering to provide amazing cruise experiences for travelers across the world.

TIMELINE

1844 The first passenger tours set sail from Southampton, England. These are considered to be the oldest recorded cruises.

1889 Electric lights are introduced to ships through the SS *Valetta*.

1900 The first ship built exclusively for luxury cruising, *Prinzessin Victoria Luise*, is launched.

1912 The mighty *Titanic* sinks to the bottom of the ocean, causing the industry to improve safety regulations.

1922 A cruise ship sails all the way around the world for the first time.

1958 The first nonstop jet service to Europe takes place, ending the common practice of transatlantic travel by ocean liner.

1977 "The Love Boat," a popular TVshow that takes place on a cruise ship, debuts. This show helped cruising become a common choice for everyday people.

1998 Disney gets into the cruise business with the launch of its first ship, *Disney Magic*.

2004 The only cruise ship currently dedicated to transatlantic voyages, *Queen Mary 2*, is launched.

2018 The world's largest cruise ship, *Symphony of the Seas*, sets sail on its maiden voyage.

GLOSSARY

architects (AHR-ki-tekts): people who design buildings and supervise the way they are built

behemoth (buh-HEE-muth): something extremely large

buoyancy (BOI-uhnt-see): able to float or stay afloat

dense (dens): something that is crowded, thick, or packed tightly, having less empty space within it

diesel (DEE-zuhl): a fuel used in diesel engines that is heavier than gasoline

geolocators (jee-oh-LOH-kate-urs): objects that help find or provide the exact location of a person, place, or thing

noncombustible (nahn-kuhm-BUHS-tuh-buhl): not capable of catching fire

pistons (PIS-tuhns): disks or cylinders that move back and forth in large cylinders

propel (pruh-PEL): to push something forward

propeller (pruh-PEL-ur): a set of rotating blades that provide force to move an object through air or water

turbines (TUR-buhns): engines powered by water, steam, wind, or gas passing through the blades of wheels and making them spin

welders (WELD-ers): people who join two pieces of metal or plastic by heating them until they are soft enough to be joined together

INDEX

SHOW WHAT YOU KNOW

1. How is such a large and heavy cruise ship able to float on the ocean without sinking?

2. Why are cruise ships such a popular choice for vacationers?

3. What have engineers done to improve cruise ship technology over the years?

4. Explain how a cruise ship moves through the water.

5. How does a cruise ship move from the shipyard, where it is built, into the open water?

FURTHER READING

DK Publishing, *Cars, Trains, Ships, and Planes*, DK Children, 2015.

Mills, Teresa, *Hey Kids! Let's Visit a Cruise Ship: Fun Facts and Amazing Discoveries for Kids (Volume 2)*, Life Experiences Publishing, 2015.

Vescia, Monique, *Cruise Ship (Choose Your Own Career Adventure)*, Cherry Lake Publishing, 2016.

ABOUT THE AUTHOR

Kaitlyn Duling believes in the power of words to change hearts, minds, and, ultimately, actions. An avid reader and writer who grew up in Illinois, she now lives in Washington, D.C. She knows that knowledge of the past is the key to our future, and wants to ensure that all children and families have access to high-quality information. Kaitlyn has written more than 50 books for children and teens. You can learn more about her at www.kaitlynduling.com.

Meet The Author!
www.meetREMauthors.com

© 2019 Rourke Educational Media

All rights reserved. No part of this book may be reproduced or utilized in any form or by any means, electronic or mechanical including photocopying, recording, or by any information storage and retrieval system without permission in writing from the publisher.

www.rourkeeducationalmedia.com

PHOTO CREDITS: Cover, page 1, 11, 12, 13, 16: ©Michael Wessels; page 3: ©johnason; page 4: ©Yobro10; page 5: ©apomares; page 6: ©Feverpitched; page 6b: ©cworthy; page 7b: ©KentWeakley; page 8, 34: ©Michael Verdue; page 10a: ©DCIM; page 10b, 11b, 17, 24, 39: ©Filippo Vinardi; page 13b: ©Carnival; page 14: ©mihailomilovanovic; page 15: ©photosuit; page 18: ©JGalione; page19: ©VvoeVale; page 20, 21, 38, 42: ©Gfincantieri; page 25: ©Sam Woolford; page 26: ©LarsZahnerPhotography; page 27: ©Aitormmfoto; page 28: ©reeixit; page 30: ©eyfoto; page 32, 33: ©LOC; page 44: ©dlewis33; page 39: ©Elenarts; page 40: ©LosLarsos

Edited by: Keli Sipperley
Cover and interior design by: Rhea Magaro-Wallace

Library of Congress PCN Data

Cruise Ships / Kaitlyn Duling
(Engineering Wonders)
ISBN 978-1-64369-048-3 (hard cover)(alk.paper)
ISBN 978-1-64369-090-2 (soft cover)
ISBN 978-1-64369-195-4 (e-Book)
Library of Congress Control Number: 2018956081

Rourke Educational Media
Printed in the United States of America, North Mankato, Minnesota